IMAGES
of America

HONOLULU TELEVISION

ON THE COVER: Alfred Apaka, a popular singer from Hawaii, earned the attention of Bob Hope, Ed Sullivan, Dinah Shore, and many others during the 1950s and appeared on their television specials and shows. Apaka was a longtime regular on the *Hawaii Calls* radio show and appeared at Henry Kaiser's Hawaiian Village Hotel as its top entertainer. Apaka was a frequent guest on local television in Hawaii during the 1950s. (Courtesy of KITV.)

IMAGES
of America

HONOLULU TELEVISION

A.J. McWhorter

ARCADIA
PUBLISHING

ISBN 978-1-4671-2758-5

Published by Arcadia Publishing
Charleston, South Carolina

Printed in the United States of America

Library of Congress Control Number: 2017941573

For all general information, please contact Arcadia Publishing:
Telephone 843-853-2070
Fax 843-853-0044
E-mail sales@arcadiapublishing.com
For customer service and orders:
Toll-Free 1-888-313-2665

Visit us on the Internet at www.arcadiapublishing.com

Honolulu Television *is dedicated to my family, who have always supported my goals and dreams over the years.*

Contents

ACKNOWLEDGMENTS

Honolulu Television would not be possible without the help of many of my friends, news colleagues, and the general managers of KITV (Joe McNamara, 2015–2017), *Hawaii News Now* (Rick Blangiardi), KHON (Kristina Lockwood) and PBS Hawaii (Leslie Wilcox). *Honolulu Star-Advertiser* photograph editor George Lee has been a tremendous help for me not only for this book but also for my *Honolulu Star Bulletin* and *Honolulu Star-Advertiser* newspaper column I wrote for a decade. That column paid tribute to people who worked in Honolulu media and made an impact over the years whether in radio, television, or entertainment. Former local sports anchors Bob Basso (KHON), Mel Proctor (KGMB), and Greg Lucas (KITV) have written books and helped inspire me to write one of my own. Mahalo! The following people and organizations also helped make this book possible with an abundance of information and photographs over the years: Wayne Collins, Ted Shibuya, Joe Konno, George Hooper, Diane Ako, Doug Bruckner, DC Video, Tom McWilliams, Paul Fox, Jim Manke, Don Robbs, Anna Gomes, Bob Basso, Emme Tomimbang, BJ Sams, *Honolulu Star-Advertiser*, KHON, KITV, *Hawaii News Now*, PBS Hawaii, Carl Hebenstreit, Florence "Johnny" Frisbie, Haumea Ho, Janet Zappala, Susan Hutchison, Roberta Wong, Paul Guanzon, Jacque Scott, Adrienne Sweeney, Marty Chase, Paul Valenti, Carmel Campbell, Julie Lauster and Pake Zane of Antique Alley, Lynne Waters, Tim Tindall, Don Rockwell, Chapman Lam, Linda Ryan, the Hawaii State Archives staff, Gary Sprinkle, Pamela Young, Don Picken, Bob Jones, Phil Arnone, the Hawaii State Library staff, and Maui's Bill Atkinson from the 50th State Big Time Wrestling website. The following friends and mentors also helped me tremendously over the years and are sadly no longer with us: Bob Sevey, Jack Kellner, John Kernell, Uncle Tom Moffatt, Bob Miller, Les Keiter, Eddie Sherman, Bill Baist, and Ron Cooper.Special thanks go to Mike Kinsella, Henry Clougherty, and Liz Gurley from Arcadia Publishing for all of their help and assistance in this book from its beginning to its fruition.

All images in this work appear courtesy of the author unless otherwise credited.

INTRODUCTION

In the early 1950s, radio was still the most dominant medium for people's entertainment and information. However, that changed with the arrival of television. When television was first launched in the islands in 1952, Hawaii was still a territory of the United States. At the beginning, local stations experienced complications, such as equipment delays. Also, the varying terrain in Hawaii proved difficult for local broadcasters. For broadcasters in Honolulu to reach other parts of the island of Oahu, let alone neighboring islands, they had to spend additional money on equipment to strengthen their signals.

It was a race to see which station could be on the air first in the islands. KONA Channel 11 was the first local station to air images on television with its test patterns and segments of *March of Time*, which were shown to the handful of people who actually owned television sets. Many island residents had purchased television sets, but a weeks-long dock workers' strike delayed many from receiving theirs.

On December 1, 1952, KGMB became the first station in the islands to air regular programming, and Kini Popo started it all off. KONA had to shut down operations for several months to fix problems with radio signal interference; at the time, KONA was sharing the same studios with KPOA radio.

When KULA came aboard in 1954 as the ABC network affiliate, island television viewers could now see all the networks' programs, instead of the other stations sharing network programs. During the early years of television in Hawaii, Arthur Godfrey's variety show was one of the locals' favorites, and island *keiki* (children) could watch cowboys and lovable hosts on locally produced shows present movies and cartoons for hours. By the time statehood rolled around in 1959, Lloyd Bridges's *Sea Hunt* was the most watched show in the islands.

Television viewers in the 50th state had an array of programs from professional wrestling, cooking shows, and hosted television movies to sports, news, variety shows, ethnic-based programs, and music dancing shows. Local shows had no problems booking national celebrity guests, as many were in town to promote themselves and enjoy the sun and fun in the islands.

Television hosts Lucky Luck, Sheriff Ken, Captain Honolulu, and others became local celebrities and made appearances at store openings, parades, and public gatherings to hundreds of their fans, who would flock to see them in order to get a photograph, an autograph, or maybe a free prize.

When *Checkers & Pogo* debuted in 1967, it faced competition from the popular *Captain Honolulu*, which had been on the air for a decade at that point. Within two years, *Captain Honolulu* had left the airwaves, and *Checkers & Pogo* with its early morning airing, coupled with the afternoon slot, had island keiki flocking home to watch. Despite valiant efforts on the part of other stations, no other kids' show could compete with *Checkers & Pogo* until the mid-1970s, when a new show from Japan, *Kikaida*, aired on KIKU. At the time, superheroes, coming alive from comic books and other media, dominated the television landscape. By the time *Checkers & Pogo* went off the air in 1982, island children who had grown up with the show were now grown up themselves and had moved on to other interests in music and video games.

Technology and transportation advancements helped bring the islands closer to the mainland United States by the 1970s. The advent of satellite communications helped island residents view far off people, places, and things. National network news was flown in on jumbo jets after West Coast broadcasts in Los Angeles. The difference in time to the mainland gave Hawaii news stations more time to create and produce television news.

KGMB newsman Bob Sevey became the most watched and trusted of all news anchors in Hawaii. KGMB news aired back to back with Walter Cronkite's *CBS Evening News*, and many dubbed Sevey "the Walter Cronkite of the Pacific." The powerful and talented KGMB on-air news and sports staff, consisting also of Tim Tindall, Bob Jones, and Joe Moore, was simply unstoppable—no other local news station could compete with that lineup.

When Joe Moore departed KGMB for KHON in 1978, followed by Tim Tindall to KITV in 1982, the familiar faces were now spread over all the news channels, and KGMB's ratings dominance began to erode slightly.

By the mid-1980s, Joe Moore, now a KHON news anchor, had taken over the top-rated news in town. Hawaii television viewers also had additional channels via cable that changed the local ratings landscape.

Television in Hawaii by 1985 had finally caught up to the mainland in terms of same-day programming as satellite costs came down. Island viewers could now finally watch network programs on the same day and not have to wait a week to see their favorite shows. Sporting events and network news were also in line on the same day, and Hawaii television viewers could see exactly what mainland television watchers saw.

KFVE joined the dial in 1988 and offered sports, movies, and syndicated television shows. The original television stations in the islands, KGMB, KHON, and KITV, have all moved from their original studios to modern facilities in large buildings.

The local media landscape changed significantly in 2009 when KHNL and KGMB merged news operations to form *Hawaii News Now*. With cable and streaming options came more channels and more choices, making television from the early days seem in the far distant past. But the people who grew up watching television in Hawaii during the 1950s through 1980s often harken back to their *hanabata* (childhood) days with fond memories of days gone by. *Honolulu Television* will take you back to those early days and show you when television in the islands was a new sensation to be seen together with families and friends.

One

The Early Years of Honolulu Television

On March 15, 1930, KGMB radio was launched from the Royal Hawaiian Hotel. Over the next decade, the station would move its studios several times, including once to the site of what is now the McKinley Car Wash at 1139 Kapiolani Boulevard. On November 16, 1940, KGMB radio settled into new studios, pictured here, on 1534 Kapiolani Boulevard, and with the expansion of the building to welcome television in 1952, KGMB would remain there until 2009. (Courtesy of Hawaii State Archives/Holst and Male Inc.)

The key staff who helped put KGMB television on the air for the first time are, from left to right, (first row) KGMB chief engineer Ernest G. Linderman, KGMB president J. Howard Worrall, and KGMB vice president and general manager C. Richard Evans; (second row) sales manager Wayne Kearl and program director Robert Costa. (Courtesy of the *Honolulu Star-Advertiser* Archives.)

As KGMB television was preparing to go on the air, the station had to test its equipment often. In this photograph, from left to right, J.B. Dearing, RCA field engineer; Dan Hunter, KGMB's chief engineer; and Julius M. Vetter, transmitter engineer, prepare test patterns to be sent out a couple weeks prior to their initial launch on December 1, 1952. (Courtesy of the *Honolulu Star-Advertiser* Archives.)

Here is a look inside the new television studios of KGMB. At the start of television, many radio performers transitioned to the new medium. Local television studios were small, and cameras were simply placed in front of background sets. (Courtesy of the Hawaii State Archives.)

The very first local television celebrity in the islands was KGMB radio announcer Kini Popo. "Hello, everybody, welcome to the first official broadcast of KGMB-television" were the first words spoken by Popo to be broadcast on island television for KGMB in 1952. Popo would host the original *Sunrise* program as well as other shows for KGMB and KONA television over the years. (Courtesy of Carl Hebenstreit.)

A large crowd gathers at the Kapiolani Furniture Store at 3654 Waialae Avenue in Kaimuki to view KGMB's first broadcast on Monday, December 1, 1952. KGMB went on the air live at 5:04 p.m. After interviews and introductions, a Gene Autry film aired at 5:30 p.m., a children's program *Time for Beanie* was next at 6:00 p.m., and Betty Smyser made a live introduction of the *Lilli Palmer Show* at 6:15 p.m. Love's Bakery was the first ever sponsor on local television and paid for a Hopalong Cassidy film. After that, the movie *Meet John Doe* capped off the initial broadcast for the station as it signed off the air at 9:43 p.m. With several hundred television sets in local television dealers' stores and in homes, it was estimated that possibly 10,000 or more local residents saw the first broadcast of television in the islands. (Courtesy of the *Honolulu Star-Advertiser* Archives.)

Top executives of KGMB are in the engineer room in November 1953 to throw switches for the station's new high-power television transmitters. From left to right are Wayne Kearl, assistant general manager; C. Richard Evans, vice president and general manager; J. Howard Worrall, president; and Daniel Hunter, chief engineer. The new transmitter cost more than $100,000 to purchase and install, and could radiate 107,000 watts of power. (Courtesy of the *Honolulu Star-Advertiser* Archives.)

Frank Valenti came to Hawaii in 1948 to work for KULA radio setting up the sports and news departments. In 1952, Valenti signed on as KGMB sports director, becoming the first sports anchor on local television in Hawaii. He would remain at KGMB for 16 years; he hosted the well-liked *Sports Editor* every Saturday night with guests ranging from Arnold Palmer, Jack Nicklaus, Sugar Ray Robinson, Elgin Baylor, and Jerry West. Another popular sports show Valenti hosted was *Biddy Boxing*, kids boxing, which was sponsored by the Police Athletic League. He was named Hawaii Sportscaster of the Year in 1961, 1964, and 1965. (Courtesy of Ted Shibuya.)

KGMB's first foreign language telecast was the *Japanese TV Revue*, which debuted on Sunday, December 21, 1952. The series is believed to be the very first Japanese language television show in American television history. Peggy Ohta and Kanichi Murata produced and hosted the show.

In this 1953 photograph, C. Richard Evans (far left) and J. Howard Worrall (center) are with Cornelio R. Gorospe, head of Philippine programming at KGMB. In the early days of local radio and television in the islands, programs were produced to help represent some of the local ethnicities in the islands. (Courtesy of the Hawaii State Archives.)

On January 19, 1953, the *Pan American World News* made its debut with Bob Jensen narrating the program. After Jensen departed to manage a local radio station, radio broadcaster Wayne Collins was hired as news anchor and news director for the station. Collins made his debut on September 16, 1953. (Courtesy of the *Honolulu Star-Advertiser* Archives.)

In the days before teleprompters, Wayne Collins memorized his own script. Collins would anchor KGMB news for the station on and off for nearly a decade. Pan American Airlines was the sponsor for KGMB news in its early days. (Courtesy of Wayne Collins.)

On November 17, 1952, KONA broadcasted the first images, scenes from the newsreel series *March of Time*, on local television in Hawaii. The station's lone network affiliate at the time was the DuMont network as KGMB had signed legal rights to broadcast ABC, CBS, and NBC programs. The first significant news program to air on KONA was US president-elect Dwight Eisenhower's arrival in Honolulu via ship. Regular programming began on December 15, 1952. KONA television initially operated out of KPOA's radio station, but there was too much interference from the radio waves. Eventually, the station had to shut down operations for several months in order to get technical difficulties taken care of and new studio operations set up.

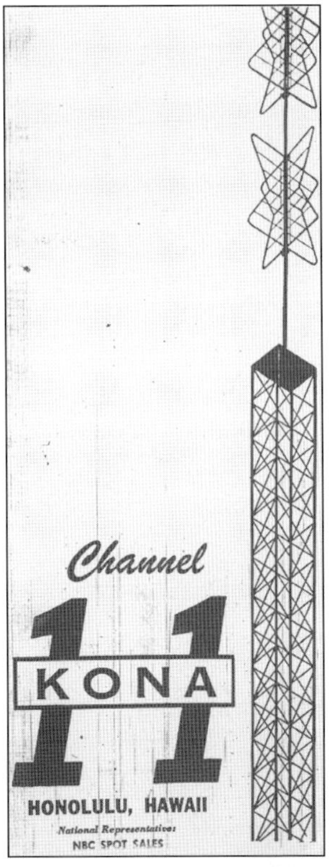

Arriving from San Francisco is the new General Electric six-bay transmitter antenna swinging over the side of a Hawaiian refiner at Pier No. 31 on April 21, 1953. The 102-pound antenna was installed on top of an existing tower at Piikoi Street and Ala Moana Boulevard. It gave KONA 34,000 watts of transmitting power. After mounting was completed, the height of the new tower was 362 feet. Prior to the antenna's installation, KONA was transmitting from a single-wing antenna with only 5,000 watts. (Courtesy of the *Honolulu Star-Advertiser* Archives.)

John D. Keating, president and general manager of KONA television, along with treasurer Allan J. McGuire, oversees installation of the new transmitter for the station. On June 24, 1953, KONA officially returned to the local broadcast airwaves at its new 5,500-square-foot building at 206 Koula Street in Kakaako. The station would now be the NBC network affiliate for the islands. (Courtesy of the *Honolulu Star-Advertiser* Archives.)

Here is a look inside the new KONA studios on Koula Street. The set of Lucky Luck's show is pictured. (Courtesy of the Hawaii State Archives.)

Lucky Luck was one of the most beloved island entertainers on both radio and television during the 1950s and 1960s. At one time, he had a radio show on KGU and three separate television shows on the air that were a mix of comedy, variety, music, and cartoons to entertain adults and kids. When KONA (now KHON) television had its official launch in 1953, Luck was among the first entertainers on the air for the station. (Courtesy of George Hooper.)

Carlos Rivas hosted *Sportsreel* for KONA television. Rivas was well known in the islands for his work in radio as a sports broadcaster for KGU. His popular recreations of baseball games during the 1950s helped bring a game broadcast thousands of miles away from Hawaii seem a little bit closer. (Courtesy of the *Honolulu Star-Advertiser* Archives.)

In 1953, KULA radio leased 26,000 square feet of land at 1290 Ala Moana Boulevard, then owned by the Victoria Ward Estate. Construction for the new television studios began in December 1953. (Courtesy of KITV.)

The station officially signed on the air on April 16, 1954. KULA would become the ABC affiliate for the islands. (Courtesy of KITV.)

In July 1954, the first morning television series in the islands, *Sunrise*, debuted on KGMB. For nearly three hours each morning, Kini Popo, along with his cohost Peaches, a two-year-old chimpanzee, covered a variety of news, weather, music, cooking, and health topics. *Shadow Stumpers*, a quiz portion of the show, generated some 60,000 calls into the station per question at one time. (Courtesy of Carl Hebenstreit.)

Kini Popo would also host shows on KONA television. Here, he is hosting the *Kini Popo Show* around the time Hawaii had just been approved for statehood in 1959. (Courtesy of Carl Hebenstreit.)

The staff of KULA television pose for a photograph during the holidays in 1955. Station manager Art Sprinkle (seated at center with legs crossed) and program director Bob Sevey (second row, far left) arrived from Phoenix, Arizona, the year before to help put the new ABC affiliate in the islands on the air. (Courtesy of Joe Konno.)

KONA television's new 121-foot, six-bay antenna was placed atop a 249-foot tower at South Queen and Auahi Streets. The new transmitting equipment doubled the station's power from 30,000 to 60,000 watts. (Courtesy of the *Honolulu Star-Advertiser* Archives.)

The new KONA antenna was placed in preparation as the station switched from Channel 11 to Channel 2. On October 22, 1955, KONA officially became the first station on the dial, Channel 2. (Courtesy of the *Honolulu Star-Advertiser* Archives.)

On May 5, 1957, Henry J. Kaiser launched his own television station on Channel 13 with the call letters KHVH, which stood for Kaiser's Hawaiian Village Hotel in Waikiki. Kaiser lived in Hawaii Kai, and for several years complained about having poor television reception there. He wanted to begin owning a series of television stations throughout the United States as part of the eventual Kaiser Broadcasting system. KHVH initially was Honolulu's first syndicated local television station. A year later, on May 7, 1958, Kaiser purchased KULA television for $685,000 and merged the two stations. Kaiser changed KULA's call letters to his own KHVH and acquired the rights to KULA's ABC network affiliation. (Courtesy of George Hooper.)

KHVH hired KGMB radio broadcaster John Needham to be the first news anchor for the station. The *Shell News* ran for 15 minutes a day with news, film, and weather. KHVH's news sponsor was the Shell Oil Company. (Courtesy of George Hooper.)

At the time, KHVH broadcast Bugs Bunny cartoons at 5:30 p.m., the *Shell News* with Needham at 5:45 p.m., and the 30-minute *Captain Honolulu* show with Popeye cartoons at 6:00 p.m. This photograph is dated March 6, 1959. (Courtesy of George Hooper.)

John Galbraith replaced John Needham as KHVH news anchor beginning in October 1959. Galbraith had nearly 20 years of broadcast experience at the time, working for both NBC and ABC networks. Initially, Galbraith anchored the *Shell News*, but a news sponsorship change to Hawaiian Electric was established the following year. (Courtesy of George Hooper.)

Joan "Becky" Collins was the first female television news broadcaster in the islands. Pictured with KGMB photographer Ted Shibuya, Collins reported the news as well as anchored a brief five-minute midday newscast from 1960 to 1961 for KGMB news. (Courtesy of Ted Shibuya.)

At the time, Becky Collins was the wife of Wayne Collins, KGMB news anchor. Collins would go on to run a public relations firm in the islands. (Courtesy of the *Honolulu Star-Advertiser* Archives.)

Joe Konno is pictured with Vice Pres. Richard Nixon in 1960. One of the first photographers hired by KULA, Konno worked for the station in its transition to KHVH before eventually working at PBS Hawaii until his retirement in the 1980s. Vice President Nixon was in Hawaii for a campaign stop as he was running for president that year. (Courtesy of Joe Konno.)

Vice-presidential candidate Spiro Agnew is pictured at the KHON studios during a campaign visit in September 1968. Agnew and Richard Nixon would win the 1968 election and become vice president and president of the United States. (Courtesy of the *Honolulu Star-Advertiser* Archives.)

In 1958, KONA television moved into its $150,000 new 10,000-square-foot studios on 1170 Auahi Street. The station would remain there until 1999, when it moved into its present location on Piikoi Street. (Courtesy of the Hawaii State Archives.)

This January 1961 photograph, taken in the KGMB newsroom, shows sports anchor Frank Valenti and news anchors Peter Burns, Wayne Collins, and Bob Miller. At the time, KGMB had the largest news staff and was owned by the *Honolulu Star Bulletin* newspaper. Collins often ended his newscast with an editorial. (Courtesy of the *Honolulu Star-Advertiser* Archives.)

Peter Burns anchored the KGMB news in the late 1950s and early 1960s. Burns came to Hawaii as a student attending the University of Hawaii. After leaving the islands, he would go on to have a long broadcasting career for both CBS and NBC news. (Courtesy of the *Honolulu Star-Advertiser* Archives.)

On September 1, 1961, KHVH became the first television station in the islands to place an emphasis on sports with *Kaiser Sports Central*, coanchored by Jim Lathrop and Gene Good. On weekends, the team of Good and Lathrop covered the local and national sports highlights along with the sporting events of the day. (Courtesy of the *Honolulu Star-Advertiser* Archives.)

Jim Lathrop had a long radio and television sports broadcasting career in the islands. Arriving in Honolulu in 1960, he first worked at KHVH and then KGMB from 1968 to 1973. (Courtesy of the *Honolulu Star-Advertiser* Archives.)

Richard Whitcomb and George Kerster coanchored the KGMB news from 1962 to 1964. Whitcomb was news director at the station and had a lengthy broadcasting career, mostly on the mainland. (Courtesy of Helene Kerster.)

George Kerster, a native of Canada, came to the islands in 1959 originally as news director of KPOA radio. He also anchored the *Pan American News* on KHVH television. (Courtesy of Helene Kerster.)

On July 4, 1962, Honolulu launched syndicated television station KTRG, Channel 13, based out of Waikiki. The station was on the air half the day and was owned by the Watamull Broadcasting Company. By 1967, the station was under new ownership by Richard Eaton, shown here, and the call letters were changed to KIKU. The daily format was changed to represent a mix of English and Japanese programming. In 1984, the call letters were changed a final time to KHNL. The station became the first FOX network affiliate in the islands and eventually switched to NBC in 1996. (Courtesy of the *Honolulu Star-Advertiser* Archives.)

Cecil Seavey had an extensive broadcasting career mostly in radio but turned to television in 1959 at KONA. Seavey was program director, news director, and eventually news anchor for the station in 1960 on the *Finance Factors News*, which was a brief five-minute recap of the day's news events, locally and nationwide. In 1962, Seavey left KONA and became the first general manager of the new syndicated station in the islands, KTRG. (Courtesy of the *Honolulu Star-Advertiser* Archives.)

John Kernell coanchored the *Matson News* on KONA television with Norman Reyes from 1962 to 1964. Later, he became Honolulu mayor Frank Fasi's director of the Office of Information and Complaints. From 1973 to 1974, Kernell anchored the KGMB *Weekend News*. (Courtesy of the *Honolulu Star-Advertiser* Archives.)

In 1964, Lawrence "Bob" Berger purchased KHVH radio and television from Henry J. Kaiser for $4 million. Berger's KHVH became a groundbreaking station with its emphasis on news, sports, and local programs. (Courtesy of the *Honolulu Star-Advertiser* Archives.)

Under Berger, KHVH became the most watched local station in the islands with the top 20 television shows and number one news team of Mason Altiery and Bob Sevey. The station was the first in the islands to broadcast its entire lineup of local and national television shows and newscasts in color. (Courtesy of KITV.)

In 1964, Cecil Heftel purchased KGMB radio and television stations for $2.5 million. During Heftel's ownership, KGMB soared to the top of the ratings in Honolulu. (Courtesy of the Heftel family.)

Heftel brought Bob Sevey back to television after a brief stint in advertising. He also moved Tim Tindall from radio to television, which helped bolster the news station into number one; KGMB stayed atop the local news ratings for nearly 15 years. (Courtesy of the *Honolulu Star-Advertiser* Archives.)

Tim Tindall arrived in the islands on June 2, 1965, and fell in love with Hawaii. He worked in radio at KGMB and moved to television news the following year.

Tindall was one of the more familiar local faces on television over the course of four decades. He worked in radio and television at both KGMB and KITV from 1965 to 1998. (Courtesy of the *Honolulu Star-Advertiser* Archives.)

On May 14, 1965, the television side of KONA was sold, and a couple of months later, the call letters were changed to KHON, which were those of a former radio station that became KPOI in 1959. Seated are John Keating with the new owners of the station DeSales Harrison and Art McCoy. (Courtesy of the *Honolulu Star-Advertiser* Archives.)

KHVH • TV — **PROGRAM SCHEDULE: SPRING / SUMMER 1965**

RED: INDICATES LOCAL PARTICIPATING PROGRAM

PROGRAM DIRECTOR
ROBERT S. SMITH

TIME	SUNDAY	MONDAY	TUESDAY	WEDNESDAY	THURSDAY	FRIDAY	SATURDAY
6:30 AM	KHVH-TV	CAPT. HONOLULU	INTRO. TO PHYSICS	PROFILE	INTR. TO PHYSICS	CAPT. HONOLULU	KHVH-TV
7:00 AM	THIS IS THE LIFE	CAPTAIN HONOLULU FAVORITES					BIG PICTURE
7:30 AM	FAITH FOR TODAY						CAPT. HONOLULU FAVORITES
8:00 AM	SUNDAY SCHOOL ADVENTURE						
8:30 AM		MY LITTLE MARGIE					HOPPITY HOOPER
9:00 AM	SUNDAY OPEN HOUSE PREVIEW	REBUS GAME					BULLWINKLE
9:30 AM	AMERICANS AT WORK/MAUI NEWS	PRICE IS RIGHT					BEANY & CECIL
10:00 AM	ALOHA STATE FARMER	DONNA REED SHOW					CASPER CARTOONS
10:30 AM	DISCOVERY	FATHER KNOWS BEST					AMERICAN BANDSTAND
11:00 AM	ISSUES & ANSWERS	FLAME IN THE WIND					
11:30 AM	KHVH SPORTS CENTRAL	LEONARD LEIGH—MUSIC BY REQUEST					KHVH SPORTS CENTRAL
12 Noon		GENERAL HOSPITAL					
12:30 PM		YOUNG MARRIEDS					
1:00 PM		ISLAND PLAYHOUSE					
2:30 PM							
3:30 PM		TRAILMASTER					
4:00 PM		CAPTAIN HONOLULU					
4:30 PM	FABULOUS FEATURE						50TH STATE THEATER
5:00 PM		LLOYD THAXTON					
6:00 PM		6 O'CLOCK REPORT					
6:30 PM	WAGON TRAIN	VOYAGE TO THE BOTTOM OF THE SEA	SCIENCE IN ACTION	OZZIE & HARRIET	JONNY QUEST	FLINTSTONES	KING FAMILY
7:00 PM			DEATH VALLEY DAYS	PATTY DUKE SHOW	DONNA REED	Farmer's Daughter Ltc 6/25 Richard Diamond Ftc. 7/2	
7:30 PM	BROADSIDE	NO TIME FOR SERGEANTS	McHALE'S NAVY	SHINDIG	MY THREE SONS	ADDAMS FAMILY	LAWRENCE WELK
8:00 PM	THE SUNDAY NIGHT MOVIE	WENDY & ME	THE TYCOON		BEWITCHED	VALENTINE'S DAY	
8:30 PM		BING CROSBY Ltc 6/21 Farmer's Daughter Ftc. 6/28	PEYTON PLACE I	BURKE'S LAW	PEYTON PLACE II	PEYTON PLACE III Ftc. 7/2 Richard Diamond Ltc 6/25	HOLLYWOOD PALACE
9:00 PM		BEN CASEY	THE FUGITIVE		COMBAT	12 O'CLOCK HIGH	
9:30 PM				JOHNNY STACCATO			TRUE
10:00 PM	10 O'CLOCK REPORT						
10:20 PM	JIMMY DEAN	BIG MOVIE					
12 Mid							

Here is the KHVH spring/summer 1965 television program guide. Note the large time slots devoted to locally produced programs, like *Captain Honolulu*, *KHVH Sports Central* on weekends, and *50th State Theater* to name a few. (Courtesy of KITV.)

Pictured is the KHVH television studios. *Captain Honolulu*'s desk and set are on the left. In the very same studio where *Captain Honolulu* was filmed, the KHVH news desk was just a few feet away. (Courtesy of KITV.)

When Bob Sevey returned to KGMB television on July 4, 1966, as news director and anchor, the station quickly became the number one channel on the dial. With a talented and well-recognized news staff, no other news station could catch the ratings dominance of KGMB for well over a decade. (Courtesy of Jack Kellner.)

The KHVH news helicopter was a fixture for radio and television reporters to cover breaking news during the 1960s and 1970s. The helicopter could also land on water if needed. (Courtesy of KITV.)

This mid-1960s photograph shows the KHVH news helicopter flying over Waikiki. Helicopter pilot Dick Brady helped news crews cover the islands. (Courtesy of KITV.)

Longtime Honolulu news anchor Mason Altiery is pictured here at Ala Moana Beach Park during halftime of the Notre Dame–Michigan State football game on November 19, 1966. That day marked the very first live satellite broadcast from the mainland to Hawaii.

When Altiery interviewed US representative Patsy Mink, suddenly Hawaii, in terms of broadcast television, no longer seemed so isolated though located thousands of miles away from the mainland. Altiery worked in Honolulu radio and television news for several decades. (Courtesy of KITV.)

During the 1960s, news stations in small television markets had just a few members on staff. This photograph shows the KHON news team from 1967 with, from left to right, Marsha Fried (weather), Dick Desmond (news anchor), Bob Devine (news reporter), Pat Wilkins (news anchor), and Ray Sweeney (sports). (Courtesy of Pat Wilkins.)

Pat Wilkins was a longtime news anchor and reporter in Portland, Oregon, but came to the islands in 1967 to anchor the KHON news. In 1968, Wilkins returned to Oregon and worked in local television news. In the 1980s, he worked at KATU news in Portland with former FOX news host Bill O'Reilly. (Courtesy of Pat Wilkins.)

KHON sports director Ray Sweeney is pictured with, from left to right, Gary Beban, Bob Apisa, Duffy Daugherty, and Herman Wedemeyer. KHON was promoting the upcoming NBC Rose Bowl broadcast on January 1, 1968. (Both, courtesy of Adrienne Sweeney.)

The October 1968 *TV Time* cover shows the KHVH news team members, from front to back, Don Robbs, Harry Endo, and Al Michaels. After anchoring KHVH news, Robbs went on to host his own show, the *Don Robbs Show*, and worked in management for KHVH radio and PBS Hawaii. For 40 years, he was also the radio broadcaster for University of Hawaii baseball games. Endo was a familiar face on television from his work on *Hawaii Five-O* as well as appearing locally in commercials for many years for Territorial Savings Bank. Al Michaels (1968–1970) worked as the sports anchor, eventually leaving Hawaii in 1971 for the mainland where he would become one of the most watched sports broadcasters of all time calling both *Sunday Night Football* and *Monday Night Football* games, baseball playoffs, World Series, Super Bowls, NBA Finals, Olympic hockey, and other professional sports, for well over 40 years. (Courtesy of Tom McWilliams.)

Chapman Lam was the first Asian to work in local news as a reporter for KGMB in 1969. After Lam appeared on local television, the doors were opened to a wider range of ethnicities in the local television news workforce. (Courtesy of Chapman Lam.)

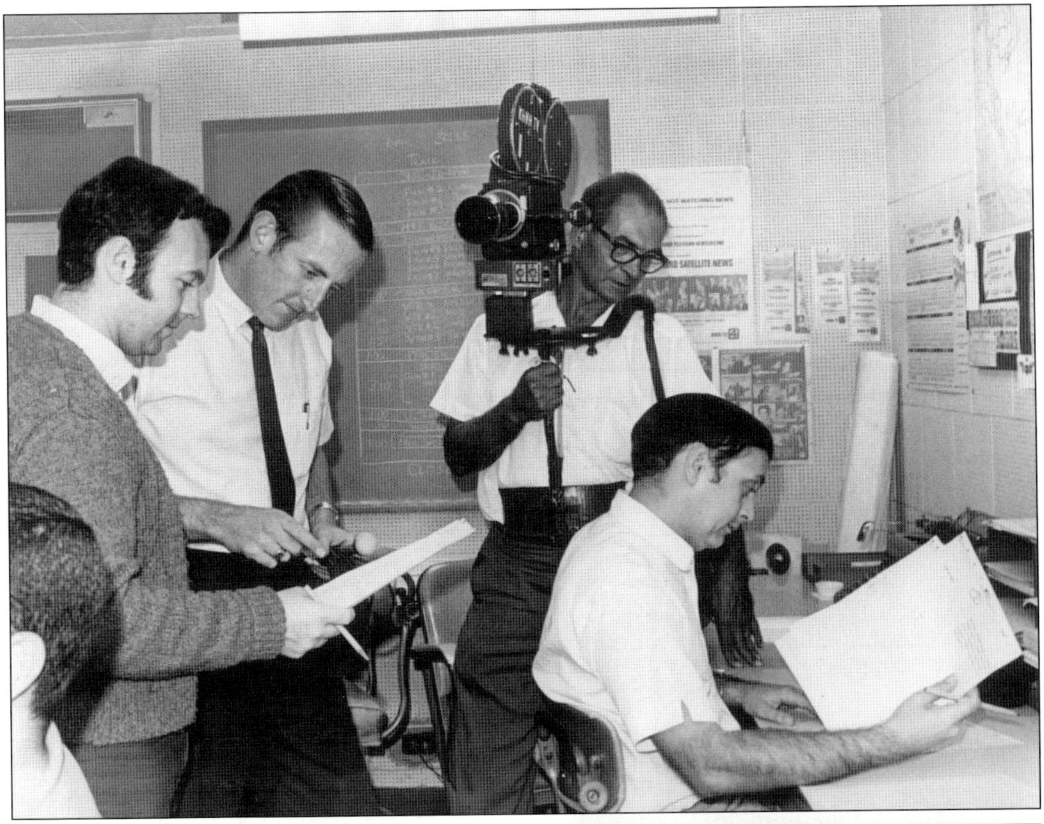

This photograph was taken in the KHVH newsroom on February 26, 1969. From left to right are KHVH news director Jim Manke, reporter Tom McWilliams, photographer George Cabral Sr., and news anchor Don Robbs. (Courtesy of the *Honolulu Star-Advertiser* Archives.)

Here, Tom McWilliams, Doug Bruckner, Don Robbs and Al Michaels are seen in a 1968 KHVH news advertisement. Doug Bruckner worked as both a reporter and anchor, eventually becoming an investigative reporter for national programs *Extra* and *A Current Affair*. (Courtesy of Tom McWilliams.)

This television ad from March 1969 promotes Chuck Henry as the newest member of the KHVH news team. Henry, just 23 years old at the time, would anchor the moon landing coverage for the station. In 1971, Henry left Honolulu for Los Angeles and then Chicago. Currently, he anchors KNBC news in Los Angeles and has worked in radio and television for 50 years.

This photograph shows the KHVH news team. From left to right and top to bottom are Don Rockwell, Chuck Henry, Al Michaels, Jim Manke, Tom McWilliams, Ken Kashiwahara, Doug Bruckner, Bart Fredo, Linda Coble, and Byron Baker. Kashiwahara coanchored the KHVH news with Henry in Honolulu as well as KABC news in Los Angeles. Kashiwahara went on to become a reporter and news correspondent for ABC news, being one of the last broadcast journalists during the fall of Saigon in 1975. Rockwell was an anchor and news director for the station from 1969 to 1985, and McWilliams was a news anchor and reporter for both KHON and KHVH/KITV.

The KHON news staff was profiled in a February 1969 *TV Time* edition. From left to right are Marsha Fried Bohnett (weather), Bob Basso (sports), John Adameck (news photographer), Don Picken (news anchor), Charles Stubblefield (news reporter), Bill Baist (news anchor), and Tommy Hisamura (news photographer).

The KHON news staff prepares for a newscast in 1969. From left to right in the foreground are Don Picken, Charles Stubblefield, Marsha Fried Bohnett, and John Adameck. (Courtesy of the *Honolulu Star-Advertiser* Archives.)

KHVH also was the first local station to broadcast election coverage in color, in 1966. Longtime Honolulu radio and television newsman Bob Miller is shown here covering the 1966 local elections. (Courtesy of KITV.)

Jack Kellner's career in local Hawaii broadcasting spanned four decades, starting in the early 1960s with KPOI Radio as a "Poi Boy" and then to KHVH News Radio. Kellner moved to television news at KGMB as a reporter, and later was assignment editor for the station. In the 1970s, he worked for Hawaii governor John Burns as director of the Office of Information and Youth Affairs. He was also the voice of Hawaii Civil Defense. In 1979, he returned to television news at KHON, working as assignment editor, news director, and the station's director for community affairs. He is pictured here interviewing Ronald and Nancy Reagan. (Courtesy of Jack Kellner.)

Two

Kids' Shows

Sheriff Ken Alford was one of the first local television personalities. Alford started in local television in 1953 with commercials on KONA's *Cisco Kid* series. He later hosted *Sheriff Ken's Posse* on KGMB in the afternoons from 1953 to 1958 and then moved over to KONA to host another show, *Sheriff Ken & His Rangers*, for another seven years. Alford tapped into the popularity of cowboys and Westerns during the 1950s and 1960s. He also was a talented musician; his Dixieland jazz group performed in Hawaii for the arrivals of Presidents Truman, Eisenhower, Johnson, and Nixon. (Courtesy of the *Honolulu Star-Advertiser* Archives.)

Chubby Roland (Roland Cloutier) was a well-received television and radio personality during the 1950s and 1960s. His show *Just Kids* aired on both radio (KGU) and television (KONA and KTRG). *Just Kids* had it all, from funny face contests, games, stunts, and prizes to cartoons, serials, comedies, and Westerns to entertain the island keiki. (Courtesy of the *Honolulu Star-Advertiser* Archives.)

Bob Smith (left) wore many hats at KULA/KHVH television from program director and promotions manager to host of several programs, including *Buckskin Bob*, *Captain Honolulu*, and the *Circle K Ranch*, as shown here in the late 1950s. Guy Madison (right) is the guest known best for playing the title role on the *Adventures of Wild Bill Hickok*. (Courtesy of KITV.)

Beginning in 1958, KHVH's Bob Smith played Sergeant Sacto on the very popular *Captain Honolulu* show during the 1950s and 1960s. The show featured Popeye cartoons, plugged commercials, skits, and audience participation. According to a survey done by the American Research Bureau, *Captain Honolulu* was the favorite daily television show for island youth in Hawaii during its time. The show aired daily at 6:00 p.m. following the KHVH news. KHVH program director Bob Smith played the character of Sergeant Sacto, who would often talk of Captain Honolulu, but the audience never saw him. At the beginning and end of each show, Sacto would peer through "spectacles" of up-turned fingers, his trademark move. *Captain Honolulu* was eventually replaced by the *Don Robbs Show* in 1969. (Courtesy of KITV.)

During the 1950s and 1960s, Sergeant Sacto, who became otherwise known as Captain Honolulu, was a favorite television character and show. Below, island keiki cannot get enough of Captain Honolulu (Bob Smith) as he makes a local appearance. (Both, courtesy of KITV.)

This early 1960s photograph shows the set of *The Lucky Luck Show* at KHVH studios. Listed on a sign at left are some of the programs that aired that day, including the *Jack McCoy Show.* (Courtesy of KITV.)

Many cities on the mainland had a *Romper Room*–like television show to educate and entertain children. In Honolulu, it was no different with Robin Mann as "Miss Robin," host of the program of the same name, which aired first on KTRG in 1964 and then on KHON for six years. Later, she hosted *Robin's Room* and *Roundabout*, a local television version of *American Bandstand*. She has been teaching sociology at Kapiolani Community College for several decades. (Courtesy of the *Honolulu Star-Advertiser* Archives.)

In 1967, KGMB debuted *Checkers & Pogo*, which became the longest-running and arguably most popular children's program in island television history. The playful antics of *Checkers & Pogo* are remembered fondly by island youth who were entertained with educational *On the Go* segments, cartoons, and prizes. The show was taped at the KGMB studios in front of a children-filled studio audience. Morgan White, who portrayed Pogo, stayed throughout the 15-year run of the program. Mr. Checkers was initially played by KGMB weatherman Jim Hawthorne and was followed by Dave Donnelly (pictured in the glasses).

The last character to play Mr. Checkers, Jim Demarest also ended up playing him the longest, from 1969 to 1978. Island keiki could watch *Checkers & Pogo* in the mornings, and the show was replayed in the afternoons on KGMB daily.

In February 1974, *Kikaida* made its debut on local television station KIKU and became very popular in the islands. During the show's heyday, *Kikaida* brought in 26 percent of the local audience, with more people watching the show than network shows *Cannon*, *Mission Impossible*, and *Chico & the Man*. Hawaii governor George Ariyoshi welcomed the cast of *Kikaida* to the state capitol on August 29, 1974. Hundreds of fans bought toys, T-shirts, and figures. (Courtesy of the *Honolulu Star-Advertiser* Archives.)

Three

LOCAL TELEVISION SHOWS

On October 23, 1954, *Filipino Fiesta* debuted on KULA television. The show soon moved over to KONA where it would remain a staple on weekends, never missing an airdate. Even when host Faustino Alejandro Respicio was ill, he made sure to have guest hosts fill in for him. Respicio wore many hats as the creator, producer, director, talent booker, and host of the show. He also would do sales for the show, including live commercials. In the era of variety shows, *Filipino Fiesta* showcased Filipino culture, entertainment, music, and news. It was the longest-running local television show in the islands and aired its final episode on August 16, 1986, after over 2,000 episodes. (Courtesy of KITV.)

During the 1950s and 1960s, the variety show format was a big hit with everyone. Locally, *Televi Digest* was a popular Sunday night program, airing for a decade live on KONA with host Barbara Lee (Barbara Kim). Celebrities like Art Linkletter, Dick Gregory, Della Reese, and Sophie Tucker were guests on the show. (Courtesy of Barbara Kim.)

Televi Digest host Barbara Lee is pictured with local special guest Teddy Tanaka. The show went off the air in 1965, after more than 500 episodes aired. It will forever be remembered as an island version of *The Ed Sullivan Show.* (Courtesy of Barbara Kim.)

Napua Stevens was a popular singer and dancer who appeared often on local television during the 1950s through 1980s. In 1957, she took over for Milli Riba to host *Gasco Kitchen*, an island cooking show on KGMB, shown on weekday afternoons at 1:30 p.m. The show eventually was called *Napua's Kitchen* and ran for eight years. Stevens also hosted her own radio shows and was a longtime cohost of the King Kamehameha parade on KITV. For her contributions to Hawaiian music, she was inducted into the Hawaiian Music Hall of Fame along with the late Don Ho in 2013. (Courtesy of the *Honolulu Star-Advertiser* Archives.)

On Sunday mornings in Honolulu for more than 20 years, Betty Smyser hosted the talk show *Conversation*, the longest-running show of its kind in Hawaiian television. First appearing on KGMB television, Smyser interviewed visiting celebrities like Frank Sinatra, Burt Lancaster, Montgomery Clift, and Marilyn Monroe, who were in Hawaii performing, filming movies, or just on vacation. In 1960, she moved over to KHVH television where she began her well-known talk show *Conversation*, interviewing local and national celebrities and politicians. In this photograph, Smyser's guest is future vice president Nelson Rockefeller. *Conversation* had a simple format with guests (usually four) sitting at a round table, conversing over coffee. The show never had a formal introduction or closing, and when it ran to 29 minutes and 30 seconds, it just ended. There were no commercials, as it was considered a public service program, and sponsorship was not permitted. *Conversation* ended its 22-year run on KHVH/KITV on December 26, 1982. (Courtesy of KITV.)

At one time, Betty Smyser had a cooking show on KHVH television. Smyser's guest on the right is actress Betty White. (Courtesy of George Hooper.)

Bob Lowrie was part of the Poi Boy deejay team from KPOI radio. Lowrie grew in popularity when he started hosting his movies on KHON in 1965, beginning with the *Beard Watchers' Movie* on Saturday nights at 5:00 p.m. and then *Beard's Movie*, which showcased horror flicks, at 10:00 p.m.

Tom Moffatt was a legendary music promoter in the islands and brought in some of the biggest entertainment acts ever. He was also a radio deejay and hosted his own music variety show on KHVH and KONA television. The *Tom Moffatt Show* had an *American Bandstand*–like format with local and mainland performers playing and kids dancing. (Courtesy of KITV.)

Kini Popo had many special celebrity guests over the years, including Elvis Presley, shown here. Nat King Cole, Debbie Reynolds, George Gobel, Ed Sullivan, Vincent Price, and Richard Nixon also appeared. (Courtesy of Carl Hebenstreit.)

50th State Wrestling, which aired on KGMB, entertained fans each week with some of the best local and national talent. The show's host was Lord "Tally Ho" James Blears, and its promoter was Ed Francis, who was also a professional wrestler. Francis began promoting local wrestling events in the islands during the late 1950s and had his own wrestling television shows on both KHVH and KGMB for many years. In this February 14, 1968, photograph, Francis is in the ring with professional wrestler "The Missing Link." (Courtesy of 50th State Big Time Wrestling, photograph by George Beppu.)

Lord "Tally Ho" James Blears (right) was a professional wrestler, promoter, and broadcaster not only in the islands but also worldwide. Blears was the main host for *50th State Wrestling*, and local fans loved his British accent. He was also acted and emceed local surfing events. (Courtesy of 50th State Big Time Wrestling.)

Honolulu Stadium was the top spot to see sporting competitions in the islands for 50 years. Professional baseball and football, high school football, rodeos, boxing matches, bullfighting, track and field meets, stock car racing, and much more was held at the stadium over the years. KHVH television filmed segments of its *Sports Central* show and aired high school sports from the stadium. (Courtesy of KITV.)

Jack McCoy was a well-known television host in Hawaii during the 1960s. At one time or another, the show bearing his name aired on every station (KHVH, KONA, KGMB, and KTRG). Audience participation, either live from the Ala Moana Center or from people watching from home, was a key component to the series. Music guests, games, and contests were popular aspects of the show as people tuned in to see what they could win on the *Jack McCoy Show.* Teenaged Linda Ryan was a regular performer on the show. (Courtesy of KITV.)

Originally hailing from England, Linda Ryan, pictured on the right in the image at left and at center below, moved to Honolulu as a teenager. She soon was broadcasting on the radio, cohosting the *Jack McCoy Show*, and later hosting her own show on KHON, *Today in Hawaii*. The talk-variety show ran several years on KHON with both local and celebrity guests including Phyllis Diller (left) and Wayne Newton (below on the right). Ryan became a successful stage, film, and television actress, appearing in many hit television shows during the 1970s and 1980s. (Both, courtesy of Linda Ryan.)

In 1969, KHVH began airing talk-variety shows called *Hawaii AM*, which aired in the mornings with host Don Robbs, and *Hawaii PM*, which aired in the afternoons with hosts Jim Becker and Al Michaels. (Both, courtesy of Tom McWilliams.)

Jim Becker (above, left) hosted the *Hawaii PM* show in 1968; here, he is with actor Sal Mineo (center), famous for his 1955 role in *Rebel Without a Cause*, and cohost Doug Mossman. *Hawaii PM* was a local talk show that aired on KHVH television and had special guests like actors, musicians, and authors. The photograph below from an episode of *Hawaii PM* shows Hawaii governor John Burns (left) and entertainer Don Ho. (Both, courtesy of KITV.)

The *Honolulu AM* show changed its name to reflect its popular host, former KHVH news anchor Don Robbs. The *Don Robbs Show* debuted in 1969 with a 90-minute format that was similar to other talk shows of its era, like *The Mike Douglas Show* and *The Merv Griffin Show*. Robbs welcomed both celebrity and local guests for five years on KHVH/KITV television. This June 9, 1972, photograph shows Don Ho greeting Robbs with a kiss; Robbs was celebrating his third anniversary on the air. (Courtesy of Tom McWilliams.)

Local radio and television personality Kani Evans hosted a show called *The Outdoor Reporter*, which aired on KHON and later KITV from the 1960s to 1980s. Evans showed the never-before-televised outdoor world in Hawaii with tales of fishing and hunting. (Courtesy of the *Honolulu Star-Advertiser* Archives.)

Bruce Carter was a columnist for the *Honolulu Advertiser* and wrote about fishing. In the early 1960s, he had his own show on KONA, *Fishing with Bruce Carter*, and in 1971, his new show *Let's Go Fishing* began on KHON and became a Sunday night staple for the station. Carter's grainy fishing films along with photographs and videos of local fishermen in the islands made the show popular for many years. Carter would soon be joined by Hari Kojima, who cohosted the show with him until June 1981. (Both, courtesy of the *Honolulu Star-Advertiser* Archives.)

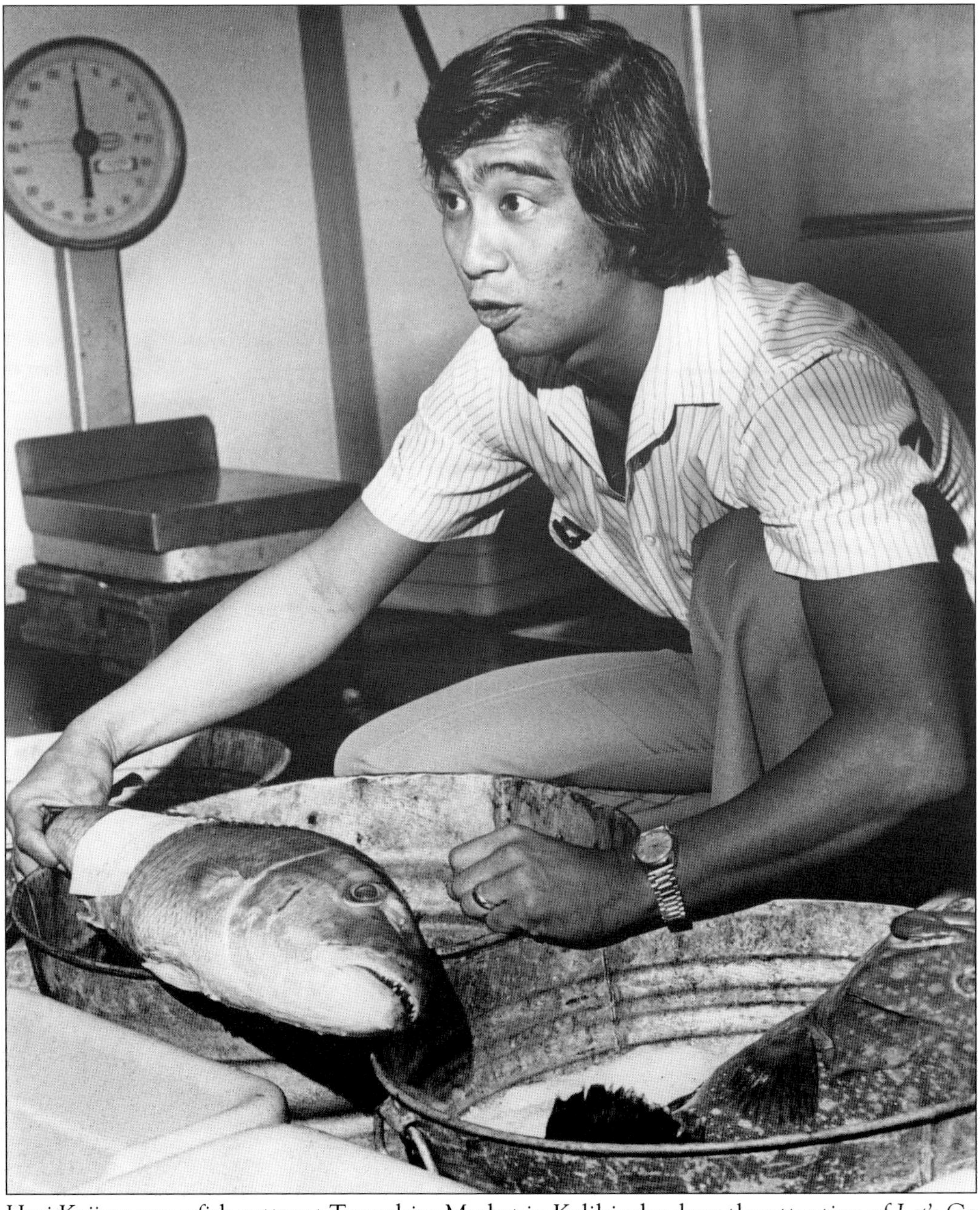

Hari Kojima was a fish cutter at Tamashiro Market in Kalihi who drew the attention of *Let's Go Fishing* host Bruce Carter. During the early 1970s, Kojima began making appearances on the show and soon became Carter's cohost. By 1981, he was cohosting the show with Stan Wright, which led to other cooking-related spinoffs for KHON until 1998. (Courtesy of the *Honolulu Star-Advertiser* Archives.)

Deejay Kamasami Kong was on the radio in Honolulu during the 1970s and 1980s. He was also on local television, hosting *Brown Bags to Stardom*, *Hawaii High*, and *Breakin' Hawaii.* He was also the original host of *Hawaiian Moving Company*. Today, Kong continues as a popular deejay in Tokyo. (Courtesy of the *Honolulu Star-Advertiser* Archives.)

Hawaiian Moving Company ran on Sunday nights on KGMB beginning in 1978 with local radio deejay Kamasami Kong as its host. At the time, the show highlighted the disco-era nightclub scene, but it evolved into showcasing the best people, places, and things of Hawaii over the course of its 25 years on the air. Michael W. Perry, pictured here facing the camera, became host in 1980, and the show became a local favorite with an estimated 3,000 segments filmed over the years with numerous talented producers including creator Phil Arnone, Larry Fleece, and Randy Brandt (far right). Also pictured are cameraman Gene Kois and crewmember Lisa Wong (seated). (Courtesy of the *Honolulu Star-Advertiser* Archives.)

The Apostolic Faith Church has been based in Kalihi for over 90 years, but its television home has been on KITV, a Sunday morning fixture, since 1979. In this photograph from 1984, KITV program director Charlie Riley and chief pastor William M. Han Sr. cut the cake to celebrate their first five years on the air. (Courtesy of KITV.)

Studio audiences were a common fixture in local television station studios across the country. Inviting guests to watch live shows helped build a strong audience base. This photograph is from the old KONA studios on Koula Street in Kakaako. (Courtesy of the Hawaii State Archives.)

Four

PBS Hawaii

This is the initial test pattern for KHET, now known as PBS Hawaii, television. KHET's call letters stood for Hawaii Educational Television. The station officially went on the air on Channel 11 with its inaugural programming on April 15, 1966, at 7:00 p.m. Channel 11 had been dormant for nearly a decade when KONA decided to leave for Channel 2 in 1955. (Courtesy of the *Honolulu Star-Advertiser* Archives.)

Bob Barker (left) and Fortunato Teho are pictured on the set of *Pau Hana Years*. Barker, a former KGMB television news anchor, was host of *Pau Hana Years* from 1966 to 1979 on KHET, PBS Hawaii. The show centered on the senior citizens of Hawaii. Teho, an agricultural expert, was producer and host of *Island Living* on KHET and had several other local radio and television programs over the years. (Courtesy of Corinne Villa-Riese.)

Master chef Titus Chan had his first local television series, *Cooking the Chan-ese Way*, debut on KHET (PBS Hawaii) in 1972. The series was syndicated nationally on some 230 PBS stations the following year. Due to Chan's popularity nationwide, he soon was appearing on talk shows such as *The Mike Douglas Show*. During the 1970s and 1980s, Chan had a television program on every local station in the islands. He became a celebrity chef and cooked for Red Skelton, Ann-Margret, Steve Lawrence, and Eydie Gorme. Today, he remains a highly requested and respected caterer in Hawaii. (Courtesy of Titus Chan.)

Titus Chan looks over thousands of fan letters in response to his cooking show on PBS. The letters came from all over the United States. (Courtesy of the *Honolulu Star-Advertiser* Archives.)

During the 1970s and 1980s, Nino Martin was a producer and director for PBS Hawaii. But he was known on air for hosting the popular *International Kitchen*, a 30-minute cooking show that showcased the diversity of cultures in Hawaii. The show aired on more than 100 PBS stations nationwide. (Courtesy of *Honolulu Star-Advertiser* Archives.)

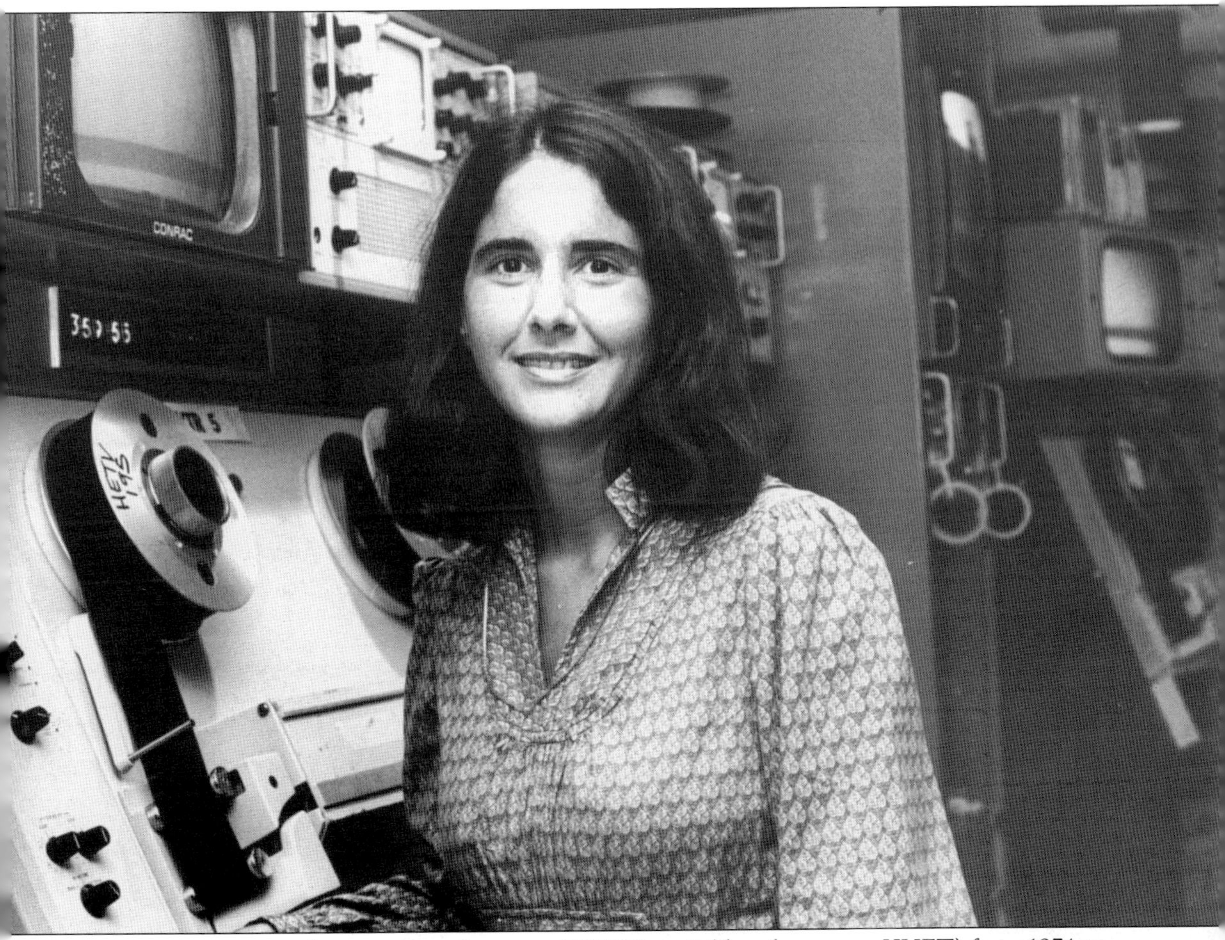

Mary Bitterman was the executive director of PBS Hawaii (then known as KHET) from 1974 to 1979. At the time, she was the youngest and only female executive director of a PBS station in the United States. During the time Bitterman ran the station, it earned a George Foster Peabody award for Aldyth Morris's *Damien*, which aired nationwide on PBS. A PBS Hawaii crew along with local residents were also among the first groups of Westerners in China when they filmed *The China Visit*. (Courtesy of *Honolulu Star-Advertiser* Archives.)

Leslie Wilcox began writing for the *Honolulu Star Bulletin* in 1973, but by the late 1970s, she was reporting for KGMB at a time when there were few female reporters in television newsrooms across the country. Wilcox quickly gained the respect of local viewers with her reports and became an anchor at both KGMB and KHON. In 2007, she became chief executive officer and president of PBS Hawaii as it transitioned into the digital age of broadcasting. Through Wilcox's leadership, the station has maintained the educational, informative, and quality programs that local viewers have always expected and come to know watching PBS Hawaii over the years. (Courtesy of PBS Hawaii.)

Five

The 1970s to Present Day

Bob Basso came to KHON news in 1968 as a sports anchor, bringing humor and crazy antics to the set every day. By 1970, Basso was news anchor and news director of the KHON *Eyewitness News* and helped create an investigative team that Honolulu television news had never seen before. In this 1972 photograph, Basso (right) interviews entertainer Bob Hope (left) and US Navy admiral John McCain Sr. Basso would later go on to have a successful acting career for several decades, appearing in top-rated network shows. (Courtesy of Bob Basso.)

On July 31, 1970, Chet Huntley retired from the long-running *Huntley-Brinkley Report* on NBC. In October of that year, KHON teamed Bob Basso (right) and Judd Hambrick (left) to form the *Hambrick/Basso Report*, which ran for over two years. (Courtesy of Bob Basso.)

Sports anchor Joe Moore (left) is seen with news anchors Bob Jones (center) and Tim Tindall in a 1971 promotional photograph for KGMB news. At the time, KGMB was the most watched news station in the islands. (Courtesy KGMB.)

KITV anchor Don Rockwell is telling viewers President Nixon will deliver his speech shortly. Television shows were often delayed a week from the mainland, but live network news was often brought in via the Lani Bird satellite. (Courtesy of KITV.)

Television trucks, which could help support being on the scene for live television, brought viewers close to the action during special live events, ranging from elections to parades. (Courtesy of KITV.)

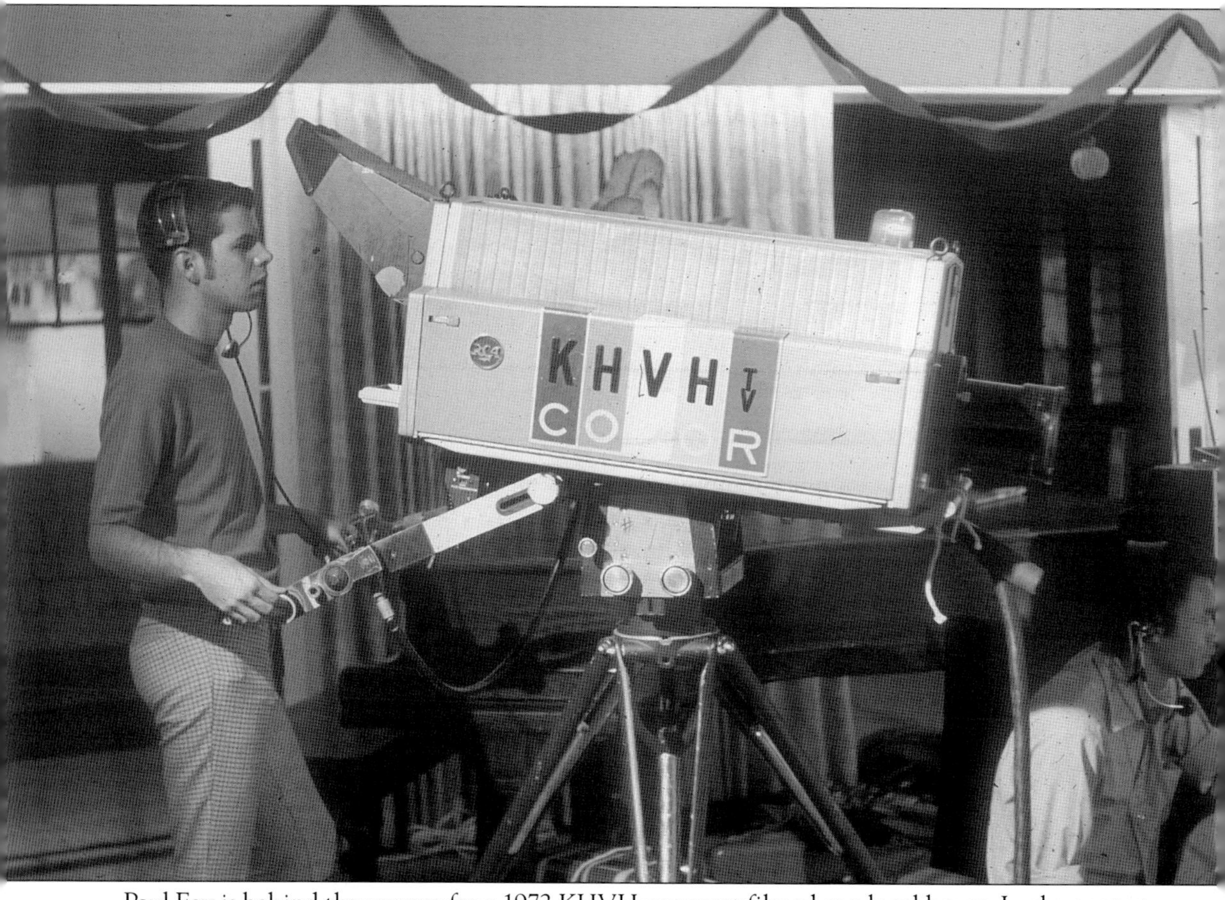

Paul Fox is behind the camera for a 1972 KHVH newscast filmed at a local house. Lucky contest winners could win a newscast in their very own home. Fox was a cameraman for the station for over 45 years. (Courtesy of KITV.)

Gangsters in Chicago? No, it is television newsmen Tom "Torpedo" McWilliams (left), "Dandy" Don Rockwell (center), and Pat "Babyface" Brown posing in downtown Honolulu for this KHVH news "Friendliest Gang in Town" promotion. (Courtesy of KITV.)

KHVH reporter Tom McWilliams and cameraman George Cabral Sr. are filming a report during the Vietnam era. The military has always been a part of the community in Hawaii, and local television news has covered the military's role on the islands. (Courtesy of Tom McWilliams.)

Mel Proctor came to Honolulu in 1973 to work at KGMB radio but was soon anchoring sports on KGMB television. In this photograph with Bob Sevey, he is calling action at a University of Hawaii (UH) basketball game. Proctor stayed in the islands until 1978, when he became the play-by-play announcer for the Washington Bullets of the NBA.

Proctor went on to broadcast the NBA finals for the Bullets and the World Series for the San Diego Padres. He has also called games for the Texas Rangers, Baltimore Orioles, and Los Angeles Clippers. He currently broadcasts UH baseball games on the radio for ESPN 1420 AM radio. (Courtesy of Mel Proctor.)

The KHON *Eyewitness News* team is pictured here. From left to right are (first row) Pete Pepper and Barbara Tanabe; (second row) Les Keiter and BJ Sams. Shortly after Sams arrived in the islands in 1973, he was in a tragic plane crash that claimed the lives of his son and Al Allen, a newsman and pilot. Island viewers welcomed Sams into their homes with an outpouring of love and admiration. He anchored the KHON *Eyewitness News* initially with Bob Basso. In 1974, Barbara Tanabe became the first woman of Asian descent to anchor the news in a local television market with a large Asian population. Tanabe coanchored the KHON news with BJ Sams for five years and later became a weekend anchor and reporter for the station, earning the respect of island viewers. Pepper, a University of Hawaii graduate, worked in radio at KCCN and became a reporter for KHON in 1973 and left to work for California television news stations in 1976. (Courtesy of Pete Pepper.)

BJ Sams (left) and Les Keiter (right) were together at the KHON news anchor desk giving the import news and sports information of the day from 1973 to 1978. Reporter and anchor Melanie Granfors (center) spent nearly a decade at KHON and also worked at KHVH radio. (Courtesy of BJ Sams.)

From left to right, KITV news anchors Warren Moran and Don Rockwell are pictured with sports anchor Greg Lucas. After Lucas left for the mainland in 1975, he went on to have a lengthy career in sports broadcasting, announcing games for the NBA's San Antonio Spurs, Dallas Mavericks, Houston Rockets, and Indiana Pacers, and Major League Baseball's Texas Rangers. Lucas is also an author and has written books on sports. (Courtesy of KITV.)

Here, KITV weekend news anchors are, from left to right, Terry Zahn, Warren Moran, and Jim Leahey. From 1974 to 1975, Zahn, Moran, and Leahey kept everyone up to date with the news on the weekends. (Courtesy of KITV.)

This 1974 photograph shows, from left to right, Don Rockwell, Warren Moran, Terry Zahn, Chuck Leahey, and Jim Leahey of KITV. Zahn was weekend anchor at the time and went on to have several anchor positions on the mainland. Moran had a long radio and broadcasting career both in Hawaii and on the mainland. The KITV sports team consisted of Chuck Leahey with son Jim Leahey. The Leahey family is well-known in the islands; for over 60 years, three Leahey generations have broadcast sports on local radio and television. (Courtesy of KITV.)

During the early 1970s, Imi Loa was a weather doll who simply pointed to the day's forecast on KITV. With his constant costume changes, he soon grew popular with both kids and adults and was once held for ransom until the on-air KITV news staff sang a song to have him released. (Courtesy of Jean Zahn.)

This September 15, 1975, picture shows the KITV *Newscenter 4* on-air staff. KITV debuted a new lineup of primarily local talent to anchor the news. After over 20 years of local television in the islands, ethnic diversity had not been seen on the local news. That changed when KITV news director Stewart Chiefet made a radical move to hire more diverse on-air talent for the news broadcasts. Other stations followed his example and hired more local talent, which can be seen on the air to this day. Pictured here are, from left to right, (sitting) Jacque Scott, Gene Okamoto, Mason Altiery, and Paul Guanzon; (standing) Don Rockwell, Roberta Wong, Matt Levi, Emme Tomimbang, Warren Moran, and Karen Ahn. (Courtesy of KITV.)

Seen here are Jacque Scott and Don Rockwell coanchoring KITV *Newscenter 4* in March 1977. Scott was a University of Hawaii graduate who transitioned from writing for the school newspaper to radio and television news at KITV. (Courtesy of KITV.)

From 1977 to 1979, Janet Zappala, Paul Guanzon (sports), and Don Rockwell anchored the KITV news together. During this time, the station experimented with having an hour-long newscast, which was unheard of in many markets in the country at the time. (Both, courtesy of KITV.)

KHON's BJ Sams (left) and Les Keiter are pictured in a relaxing moment away from the anchor desk. From 1973 to 1978, the duo helped bring the top news and sports information of the day to local viewers. As a sports broadcaster, Les Keiter called virtually every sport. His Hawaii connections dated to the late 1940s and early 1950s, when he was sports director at KPOA radio. He would return to the islands many years later, and in 1971, became KHON sports director anchoring sports and giving his *General Opinion* on sports for the station until the mid-1990s. Keiter died in 2009 and was one of the most respected and loved on-air personalities the islands have ever had. (Courtesy of BJ Sams.)

KHON anchors, from left to right, BJ Sams, Barbara Tanabe, and Joe Moore are pictured in 1978. (Courtesy of BJ Sams.)

There is no one in island television news history who can match the years of on-air experience Joe Moore has. Starting in 1969 when he first appeared on local television as a sports anchor for KGMB, Moore, along with Bob Sevey, Bob Jones, and Tim Tindall, had the island's top-rated newscasts. When Moore left for KHON in 1978, that station began a slow rise in the ratings, and eventually unseated KGMB. When Moore started anchoring the news for KHON in 1981, the ratings began to soar even higher, as he became the most trusted anchor in the islands. Moore has not only set longevity records locally but is one of the longest continuously serving news anchors at one station in the United States today. (Both, courtesy of the *Honolulu Star-Advertiser* Archives.)

The KITV weekend news crew in 1979 was, from left to right, Chris Allen, Susan Hutchison, and Bill Browning. Hutchison and Browning coanchored the news, and Allen was the sports anchor. After leaving Hawaii in 1980, Hutchison had an extensive news career as anchor in Seattle at KIRO and became chair of the Republican party for the state of Washington in 2013. Browning was a longtime radio and television announcer in California during the 1970s and 1980s. (Courtesy of KITV.)

From 1978 to 1980, the KITV sports team of Chris Allen (left) and Paul Guanzon (right) brought island viewers the top sports updates seven days a week. Allen, a University of Hawaii graduate, went on to work in Indianapolis television and, later, with the *Voice of America*. (Courtesy of KITV.)

Jim Scoutten has been reporting and hosting stories on the firearms industry for well over 20 years. His *Shooting USA* series, which airs on the Outdoor Channel, is the longest-running television series covering firearms sports and industry. Scoutten got his television start in the islands as a reporter and anchor for KITV, starting in 1978. (Courtesy of KITV.)

After six years with KHON, BJ Sams went to KITV and became its weekend news anchor in 1980. Sams (right) and Rick Quan (left) covered the weekend news for KITV together from 1980 to 1982. (Courtesy of KITV.)

KITV news director Paul Udell (center) talks with KITV news anchors Tina Shelton (right) and Tim Tindall in a 1987 photograph. Paul Udell first arrived in the islands anchoring the KHON news in 1979. As a new decade was unfolding, island news viewers saw a different style and format of news with Udell broadcasting either standing up or sitting on a couch. His news background was one of the most impressive Honolulu television news had ever seen with his work in New York, Chicago, San Francisco, and other major news markets. Udell was also a reporter and anchor at KITV from the mid-1980s through 2006. (Courtesy of the *Honolulu Star-Advertiser* Archives.)

Susan Hutchison and Jack Hawkins are coanchoring an August 16, 1980, KITV newscast. Hawkins anchored news at KITV from 1979 to 1982. (Courtesy of KITV.)

KITV news anchor Jack Hawkins is with Paul Guanzon during a lighter moment of a KITV newscast. After leaving Honolulu television news in 1982, Hawkins would later land in Nashville working in local news there. (Courtesy of KITV.)

On August 29, 1980, a 150-foot section of KHON television's antenna came crashing down. That section of the tower weighed 7,500 pounds and fell while being removed by a crane. The accident happened when the arm of the crane broke while the top portion was being removed. The tower fell into the parking lot next to the station and crushed a parked car and damaged another vehicle and nearby fence. No one was injured during the accident. (Courtesy of the *Honolulu Star-Advertiser* Archives.)

Larry Price played football for the University of Hawaii, served as an assistant coach, and eventually became the head coach during the mid-1970s. A few years later, he became a reporter for KITV news and was known for his investigative reporting and lengthy series on certain topics. By the mid-1980s, he moved to the radio broadcast booth, cohosting the *Perry & Price* morning show for over 30 years. (Courtesy of KITV.)

From left to right, Larry Price, Lynne Waters, Tim Tindall, and Rick Quan pose at the Hawaii State Capitol. They were the core on-air news talents on KITV during the early to mid-1980s. (Courtesy of Lynne Waters.)

Over his 30-plus years in local television news in Hawaii, at various times, Bob Sevey served as a program director, television show host, news anchor, news director, station manager, and yes, in some situations, also a photographer. Here, Sevey films a building fire. (Courtesy of the *Honolulu Star-Advertiser* Archives.)

Bob Sevey is relaxing in his office before an interview about the local television news ratings war in the 1980s. At the time, Sevey was competing against his former news colleagues Joe Moore at KHON and Tim Tindall at KITV. (Courtesy of the *Honolulu Star-Advertiser* Archives.)

Competition in news is one thing, but on the baseball field, it can get even more competitive. Members of the local media pose for a photograph as they prepare to square off on the baseball diamond. (Courtesy of Tom McWilliams.)

KGMB news anchor Bob Sevey had a lot to smile about during many years of local ratings dominance. He eventually retired from local news on July 4, 1986. (Courtesy of the *Honolulu Star-Advertiser* Archives.)

About 200 curious onlookers watch a May 15, 1987, KGMB newscast "on the road" at Windward Mall in Kaneohe with, from left to right, Bob Jones, Linda Coble, and Larry Beil. (Courtesy of the *Honolulu Star-Advertiser* Archives.)

The KITV news anchor team from 1982 to 1985 all originally hailed from Texas. From left to right are Rick Quan, Tim Tindall, and Lynne Waters. Quan spent seven years with the station as sports anchor and has worked in the San Francisco Bay area as a sports anchor and documentary film producer for the past 30 years. Tindall went to KITV after 17 years as anchor at KGMB. Waters became KITV news anchor in 1981, spending four years at the station as both reporter and anchor. She later became program director at KGMB and spokesperson for the University of Hawaii. (Courtesy of KITV.)

From left to right in 1985 are KITV anchors Tim Tindall, Lynne Waters, and Russell Shimooka. From 1981 to 1989, Shimooka worked as a reporter, sports anchor, and sports director for KITV. In 1989, he left to work for mainland television stations in Miami, San Francisco, and Minneapolis. Shimooka returned to Hawaii in 1997 to work as news anchor at KGMB. (Courtesy of Lynne Waters.)

Olivia Newton John is seen preparing for a speech to accept the 1983 American Music Awards Favorite Pop Female Artist award. John was in Hawaii at the time, and this segment was filmed at the KITV studios and broadcast live via satellite. (Courtesy of KITV.)

KITV's Lynne Waters and Tim Tindall coanchor the Statehood Jubilee Celebration at the Hawaii State Capitol in 1984. KITV was known in the 1970s and 1980s for covering parades, festivals, and special events to a greater extent than any other station. (Courtesy of Lynne Waters.)

After over 20 years of anchoring the news in Honolulu, Tim Tindall signs off at KITV. Emme Tomminbang signed off after 12 years of working at KITV in weather, reporting, and anchoring. Tindall returned to KGMB news in the 1990s for a few years, and Tomminbang worked at KHON news and later launched her own successful production company, Emme Inc., producing local specials. (Courtesy of KITV.)

Joe Moore and Les Keiter are seen at the KHON anchor desk in 1987. For over a decade, the duo showed respect and admiration for each other in each newscast. (Courtesy of the *Honolulu Star-Advertiser* Archives.)

These KITV news team members are, from left to right, Robert Kekalua, Dick Allgire, Paula Akana, Dan Cooke, Tina Shelton, Gary Sprinkle, Pamela Young, and Rob Fukuzaki. Kekaula was a longtime sports director at the station, and in 2017 is coanchor with Paula Akana on the 6:00 p.m. and 10:00 p.m. newscasts. Allgire reported at KGMB and anchored the news at KITV during the 1980s and 1990s. Dan Cooke was a local deejay in the islands, anchored weather at KITV, and moved to news anchoring at both KITV and KHNL. Today, Cooke anchors weather at *Hawaii News Now.* Shelton was one of the first female weekday solo anchors in the islands, anchoring at KGMB, KITV, and KHON. She currently is the communications director at John A. Burns School of Medicine at the University of Hawaii. Sprinkle was a sports anchor and sports director for KGMB news and moved to news anchoring at the station in 1986. In 1991, he moved to KITV and coanchored the news with wife Pamela Young until 2010. Sprinkle is now CEO of Sprinkle Media Solutions. Young, a news anchor at KITV, has also reported at KGMB and is currently at KHON. Her *Mixed Plate* specials have earned numerous awards over the years. *Mixed Plate* is the longest-running feature series in Hawaiian television. (Courtesy of KITV.)

Here is the KITV weekend news team. From left to right are Dalton Tanonaka, Paula Akana, and Rob Fukuzaki. Tanonaka, an award-winning journalist, has had a career based in Asia for more than 20 years and created and anchored English-language programs for Metro TV, an Indonesia news channel. He has also anchored for CNBC, CNN, and NHK. Akana has remained as the longtime face and voice of KITV for over 30 years, anchoring the flagship 6:00 p.m. and 10:00 p.m. weekday broadcasts for the station. Fukuzaki, the first male Japanese American television anchor in Los Angeles, has anchored sports for KABC since 1994. (Courtesy of KITV.)

For most of the first 40 years in Honolulu broadcast news, anchors would spend a minute or two telling viewers the weather. During the 1990s, an emphasis on weather became more of a focus on the local news stations. The late Sharie Shima began anchoring weather at KITV in the early 1990s and eventually became Hawaii's first broadcast meteorologist. She spent another eight years at KHNL, earning an Emmy Award and the respect and praise of her colleagues. (Courtesy of KITV.)

This KITV sign welcomed staff and visitors to the station during the 1980s. After 43 years at 1290 Ala Moana Boulevard, the station moved in December 1997 to One Archer Lane on 801 South King Street. (Courtesy of KITV.)

This mid-1980s photograph shows the KHON television studios and antenna. The station moved in 1999 to its new studios located on the ground floor of the Hawaiki Tower. The South Shore Market currently occupies the property that was home to KHON for over 40 years. (Courtesy of the *Honolulu Star-Advertiser* Archives.)

On October 15, 2006, a 6.7-magnitude earthquake hit the Big Island of Hawaii but was also felt on Oahu. The earthquake caused damage to many buildings and businesses, mostly on the Big Island, and a lengthy power outage on Oahu, which frustrated many residents. In some cases, local news covered the event with backup generators. In this photograph, KHNL news staff are preparing to cover the earthquake's damage. (Courtesy of Diane Ako.)

From left to right, KHON's Jai Cunningham, KITV's Jodi Leong, KHNL's Diane Ako, and KGMB's Jim Mendoza are seen covering election night in 2008. The news business may be competitive, but it also creates friendships and respect among colleagues. The last year that all four local news stations covered the elections was 2008; the following year, KHNL and KGMB merged news operations into *Hawaii News Now*. (Courtesy of Diane Ako.)

Diane Ako and Howard Dashefsky are pictured on the set of KHNL *Morning News* in 2009, the final year of news operations for the station. Ako, one of the longest-running on-air talents for KHNL, was at the station from 1996 to 2009. Dashefsky, who played on the University of Hawaii baseball team that reached the College World Series in 1980, also anchored both news and sports at local stations KGMB, KITV, and currently, KHON. (Courtesy of Diane Ako.)

There are a.m. and p.m. news crews who deliver the news to viewers on a daily basis. On Friday, October 23, 2009, KHNL's a.m. morning crew produced its final newscast. (Courtesy of Diane Ako.)

The KGMB studios are seen here during the 1950s. The building, which housed Honolulu's CBS affiliate, was also home to radio, television, professional wrestling, kids programming, and commercial production from 1940 to 2009. (Courtesy of Hawaii State Archives/Holst and Male Inc.)

This 2007 photograph shows KGMB studios at 1534 Kapiolani Boulevard. The building has been extensively renovated and is a multiuse commercial property with businesses and restaurants. (Courtesy of the *Honolulu Star-Advertiser* Archives.)

On June 25, 1962, Anita Brady started work as KGMB's receptionist. She was not just the soft and pleasant voice of the station people heard when they called, but also the gatekeeper of all things that came through the studio. She was the heart and soul of KGMB and was beloved by all. She retired on Friday, June 26, 2009, spending a remarkable 47 years with the station. (Courtesy of the *Honolulu Star-Advertiser* Archives.)

KGMB's Bob Jones, Tim Tindall, and Joe Moore anchored together at KGMB from 1969 to 1978 with Jones and Tindall handling news and Joe Moore sports. The trio would split up eventually and ended up news anchoring at different stations during the 1980s and 1990s, with Jones remaining at KGMB, Tindall at KITV, and Moore at KHON. (Courtesy of the *Honolulu Star-Advertiser* Archives.)

In October 2009, after nearly 70 years on Kapiolani Boulevard, KGMB moved its studio to Waiakamilo Road in Kalihi. The longtime home of *Checkers & Pogo* and many other "Colorful 9" KGMB news and productions may be gone, but the memories never fade away.

Due to a news operations merger with KHNL (the NBC affiliate) and KGMB (the CBS affiliate), the news station is called *Hawaii News Now*. Pictured here on Monday, October 26, 2009, Keahi Tucker and Stephanie Lum coanchor the first broadcast. (Courtesy of the *Honolulu Star-Advertiser* Archives.)

This photograph was taken looking up into the sky through the old KGMB tower. The tower remained part of the Honolulu landscape for nearly 50 years. (Courtesy of the *Honolulu Star-Advertiser* Archives.)

The familiar 432-foot KGMB television and radio tower was built in 1962 and could be seen for miles in many directions. The tower was dismantled beginning in December 2010 and ending a few months later. (Courtesy of the *Honolulu Star-Advertiser* Archives.)